CODE MONKEYS Write Algorithms

BY JOHN WOOD

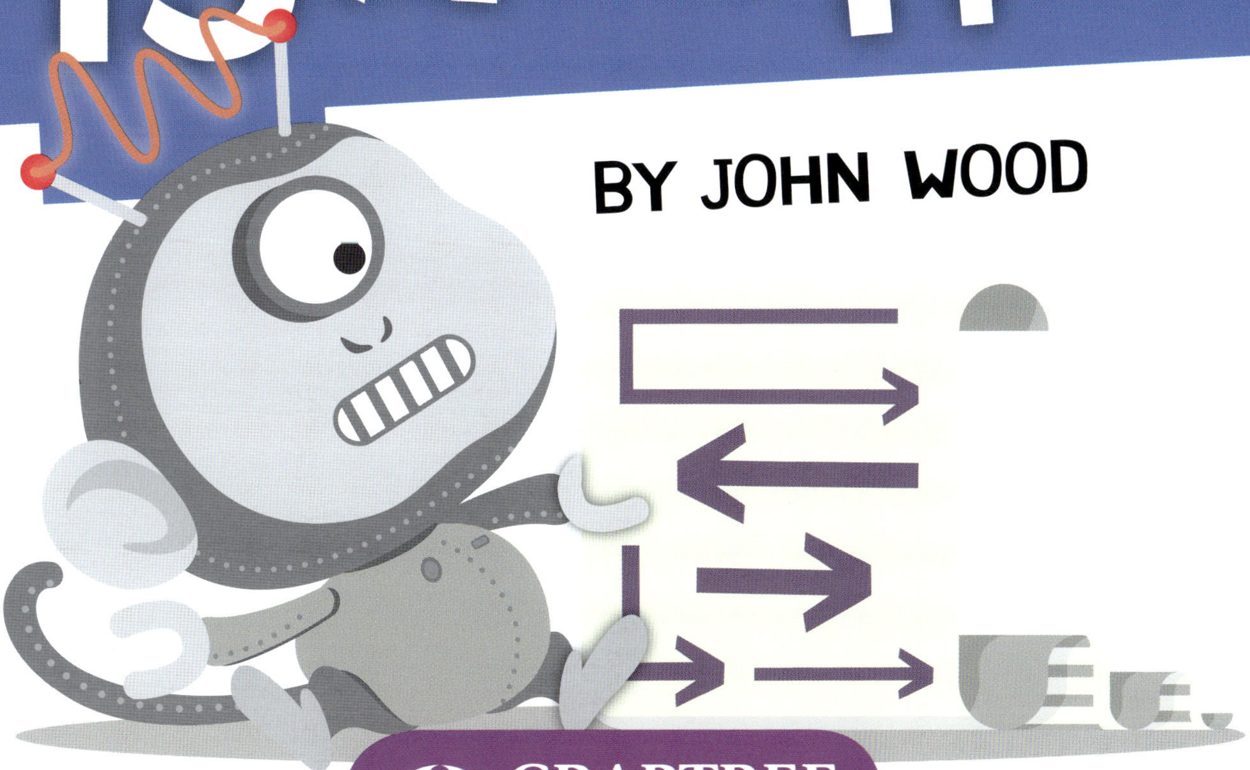

CRABTREE
PUBLISHING COMPANY
WWW.CRABTREEBOOKS.COM

CRABTREE
PUBLISHING COMPANY
WWW.CRABTREEBOOKS.COM

Author: John Wood

Editorial Director: Kathy Middleton

Editors: Madeline Tyler, Janine Deschenes

Proofreader: Melissa Boyce, Petrice Custance

Designed by: Danielle Webster-Jones

Production coordinator and Prepress technician: Margaret Amy Salter

Print coordinator: Katherine Berti

Images are courtesy of Shutterstock.com. With thanks to Getty Images, Thinkstock Photo and iStockphoto. Cover & throughout - Ori Artiste. 2 - Ann679. 4 - Ann679, GoodStudio, SK Design, feelplus, nexusby. 5 - VRstudio, cobalt88. 6 - Bloomicon. 7 - vasabii, Donnay Style, pp1. 8 - Dmitriy Domino. 9 - Diana Hlevnjak, AlexHliv. 10 - Tetiana Yurchenko. 11 - Zoran Milic, Victor Z. 13 - MSSA. 14 - Kit8.net. 15 - OHishiapply. 16-17 elenabsl, Morumotto, Andrey_Popov, Oleg and Polly, fatmawati achmad zaenuri, Alongkorn Sanguansook. 20 - cosmic_pony. 21 - MA8. 22 - curiosity. Additional illustration by Danielle Webster-Jones.

Scratch is a project of the Scratch Foundation, in collaboration with the Lifelong Kindergarten Group at the MIT Media Lab. It is available for free at https://scratch.mit.edu

Library and Achives Canada Cataloguing in Publication

Title: Code monkeys write algorithms / by John Wood.
Other titles: Algorithms
Names: Wood, John, 1990- author.
Series: Code monkeys.
Description: Series statement: Code monkeys | Originally entered under title: Algorithms. King's Lynn: BookLife Publishing, 2020. | Includes index.
Identifiers: Canadiana (print) 20200223178 | Canadiana (ebook) 20200223259 | ISBN 9780778781523 (hardcover) | ISBN 9780778781561 (softcover) | ISBN 9781427125866 (HTML)
Subjects: LCSH: Computer algorithms—Juvenile literature. | LCSH: Computer programming—Juvenile literature.
Classification: LCC QA76.9.A43 W66 2020 | DDC j005.1—dc23

Library of Congress Cataloging-in-Publication Data

Names: Wood, John, 1990- author.
Title: Code monkeys write algorithms / by John Wood.
Description: New York : Crabtree Publishing Company, 2021. | Series: Code monkeys | Includes index.
Identifiers: LCCN 2020016160 (print) | LCCN 2020016161 (ebook) | ISBN 9780778781523 (hardcover) | ISBN 9780778781561 (paperback) | ISBN 9781427125866 (ebook)
Subjects: LCSH: Computer algorithms--Juvenile literature. | Computer programming--Juvenile literature.
Classification: LCC QA76.9.A43 W66 2021 (print) | LCC QA76.9.A43 (ebook) | DDC 005.13--dc23
LC record available at https://lccn.loc.gov/2020016160
LC ebook record available at https://lccn.loc.gov/2020016161

Crabtree Publishing Company

www.crabtreebooks.com 1-800-387-7650

Published by Crabtree Publishing Company in 2021

Printed in the U.S.A./082020/CG20200601

Published in Canada
Crabtree Publishing
616 Welland Avenue
St. Catharines, Ontario
L2M 5V6

Published in the United States
Crabtree Publishing
347 Fifth Ave
Suite 1402-145
New York, NY 10016

Contents

Words with lines underneath, like this, can be found in the glossary on page 24.

Welcome TO THE JUNGLE

Code monkeys are silly and smart. They want to learn all about computers and coding. You can follow the code monkeys and learn about coding too!

A code monkey's curiosity can get it into trouble! Plug that back in!

SOME WORDS TO KNOW

COMPUTER = A machine that can carry out <u>instructions</u>.

CODING = Writing a set of instructions that tell computers what to do. The instructions are called code.

PROGRAMMER = A person who writes code, or puts instructions into a computer.

Computers are everywhere. Laptops, smartphones, desktops, and tablets are all computers. There are even computers in surprising places, such as fridges and pianos.

Coding IN THE WILD

Every game, video, and website you find on the Internet is built from code. Programmers wrote many instructions to make them work as they should.

One day, you could write code and build an amazing game!

Humans speak to each other in a language that both people understand. Code also needs to be written in a language that the computer understands. Computers understand a language called <u>binary</u>. Programmers write code in <u>programs</u> that <u>translate</u> instructions into binary.

CHIMPTUBE

If you could make a website, what would you put on it?

WHAT IS AN Algorithm?

An algorithm is a set of instructions. They tell a computer exactly how to complete a task. Programmers write algorithms very carefully.

Algorithms complete tasks and solve problems. An algorithm opens search results when a person looks for a video on a website. An algorithm solves problems such as "how will the computer count the score in this game?"

Simple algorithms complete one task, such as opening a page. More complicated algorithms complete more than one task. Google uses a complicated algorithm to open the best search results.

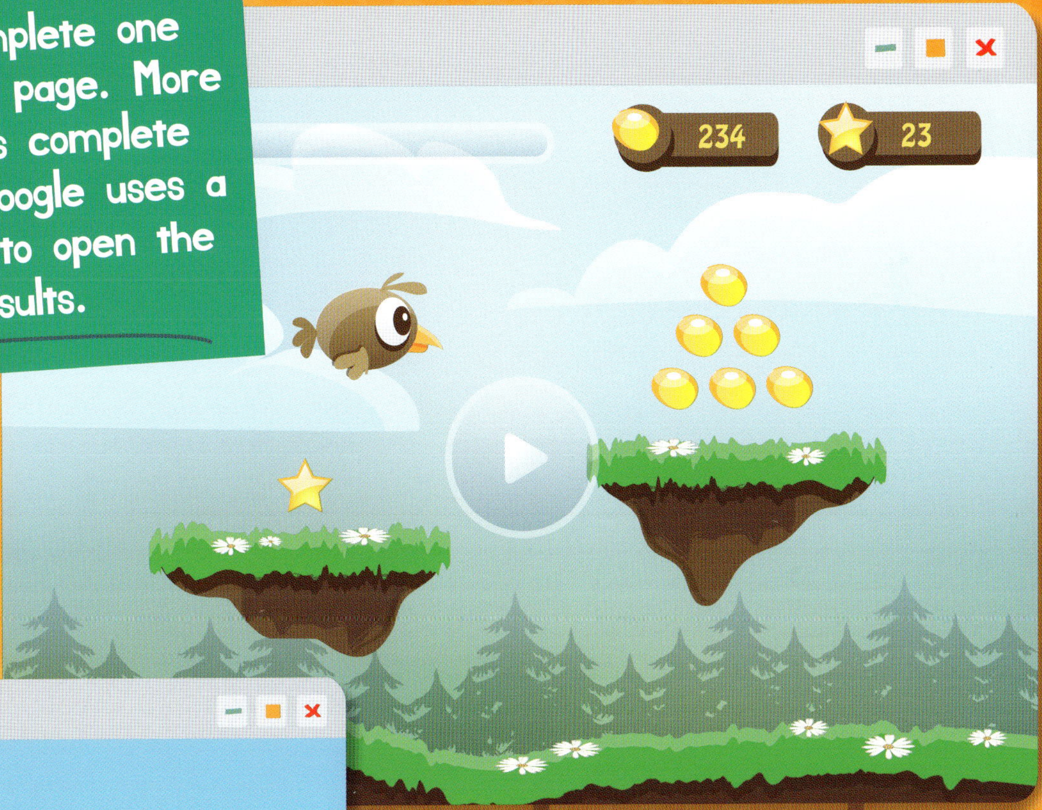

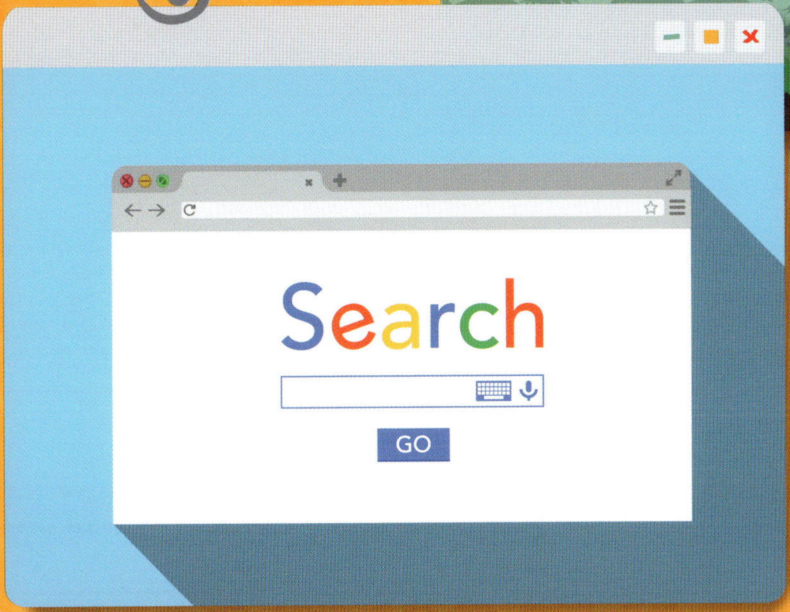

WHAT CAN Computers DO?

Computers are very good at some things. They can follow instructions and read code perfectly. They can add up numbers correctly every time. They can do the same thing over and over again without getting tired.

However, there are some things computers cannot do. They cannot come up with ideas. They are not able to solve problems on their own. To do these things, computers need simple instructions to follow.

If there is something wrong with an algorithm, the computer will not work as it should.

Flowcharts

Flowcharts can help us see how a computer follows instructions. A flowchart is a chart that shows each step of a set of instructions. It uses lines with arrows to show the correct order of the steps.

Look at the flowchart on page 13. It shows the steps that Gizmo the robot monkey will follow.

This algorithm will teach Gizmo to swing between these trees. Each step tells the robot exactly what to do.

SWING THROUGH TREES

- Jump from branch →
- Grab onto vine with two hands →
- Hold onto vine for four seconds as it swings through the air →
- Let go of vine →
- Grab onto the next branch with two hands →
- Make excited monkey noises

Programs

A program is a collection of many algorithms. Programs let computers do certain tasks. For example, to open the Internet, a computer needs to run a program called a web browser.

To run a program means that the computer goes through a program's algorithms and does what it is told.

A program is written in a programming language. There are many different programming languages. Each one uses certain words, symbols, and rules to give computers instructions.

A program needs a device on which to run. For example, this game is a program. It is run on this computer. It could also be run on a smartphone or tablet.

Algorithms IN THE REAL WORLD

Algorithms complete many important and complicated tasks in the real world.

Algorithms control the programs that help artists create animated films.

The algorithms in a GPS find the best route for a vehicle to take.

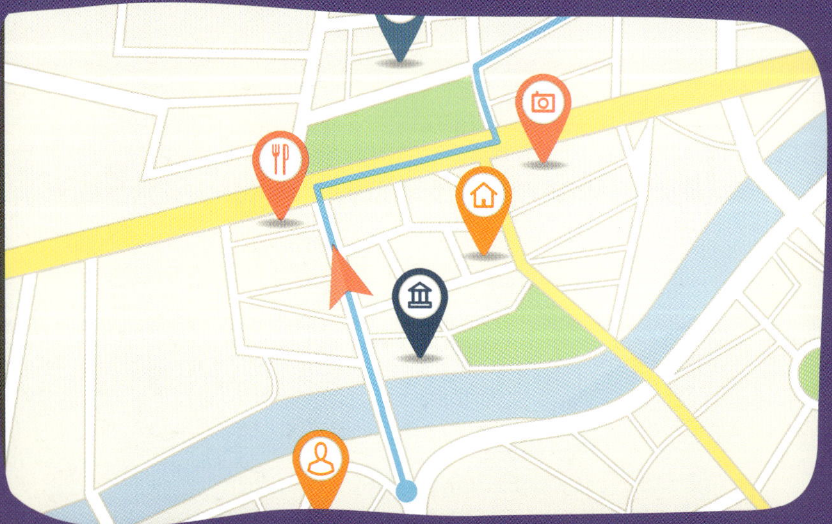

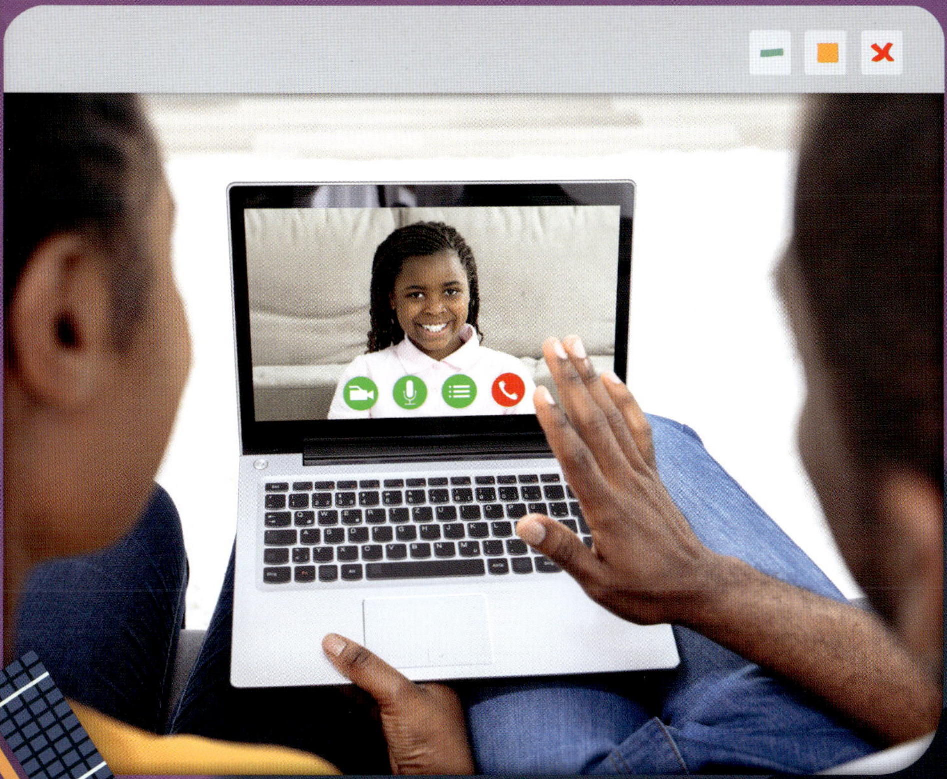

Algorithms compress sounds and videos. This lets you talk to people across the world on video calls.

Algorithms figure out how to give you the best results when you search for something on the Internet.

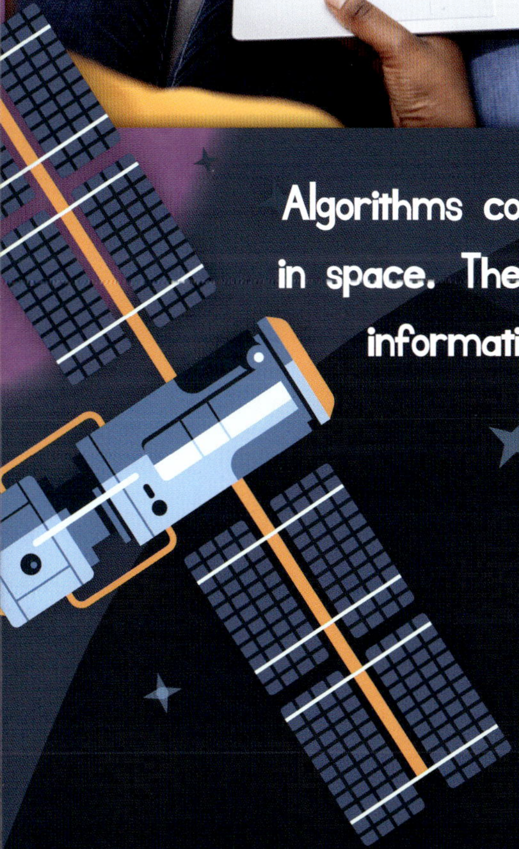

Algorithms control satellites up in space. These satellites send information to Earth.

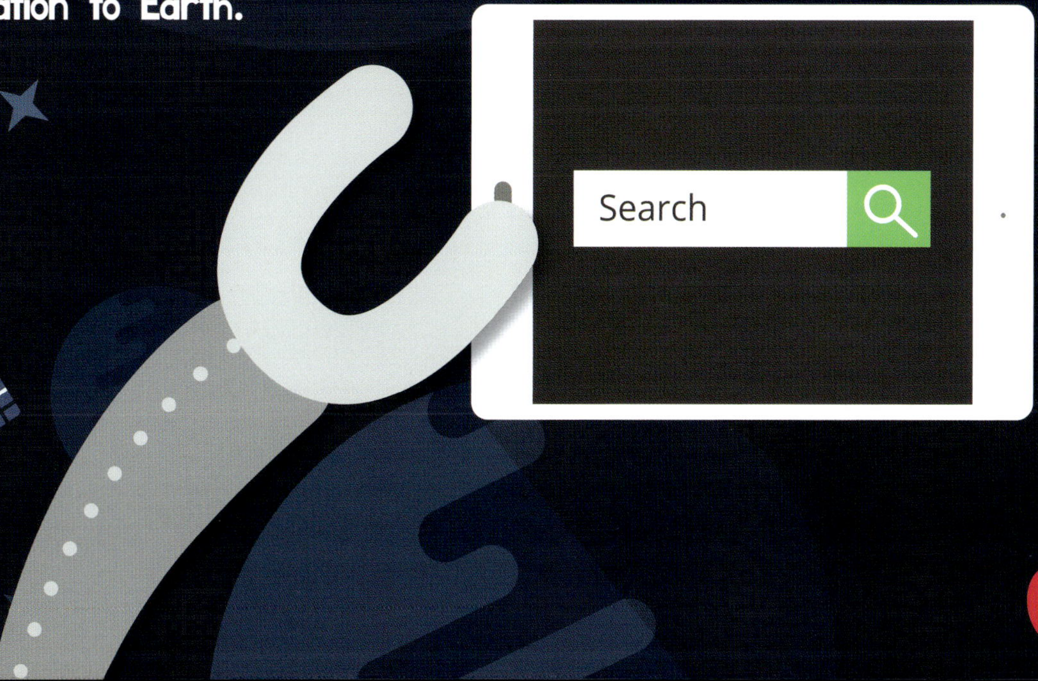

BLOCKS OF Code

Algorithms can be sorted into small blocks, or groups, of code. Each block includes two or more statements of code. Blocks of code can be given names and used more than once.

Functions are blocks of code. Each function completes a certain task. Functions can be used again and again.

Programmers can use functions instead of writing out many instructions. This makes coding faster and easier. For example, this function tells Gizmo how to pick his nose. It is called nosePick. It includes all the instructions Gizmo needs.

```
nosePick

Put finger up nose
Move finger around to collect booger
Take finger out of nose
Roll booger
Flick booger
```

Now, every time you want Gizmo to pick his nose, you can tell the computer to run the nosePick function.

Monkey See

Let's look at how an algorithm looks in the programming language Scratch. This algorithm helps Gizmo stay dry when it rains. In Scratch, the algorithm starts when the green flag is clicked.

```
when [green flag] clicked
say (It's raining!) for (2) seconds
glide (3) secs to (Tree1 ▾)
switch costume to (Monkey Sitting ▾)
```

Here is a function in the programming language <u>Python</u>. It makes Gizmo guess a number, wait for one second, and then roll a die. If Gizmo is right, he says "I win!" If Gizmo is wrong, he says "Oh dear. I'll try again" and starts the process again.

```
from random import randint
from time import sleep, time

def monkeyRoll():
    guess = str(randint(1,6))
    print('I pick the number ' + guess)
    sleep(1)
    dice = str(randint(1,6))
    print('Rolling dice... it\'s a ' + dice)
    if guess == dice:
        print('I win!')
    else:
        print('Oh dear. I\'ll try again.')
        monkeyRoll()

monkeyRoll()
```

Monkey Do

It is time to build your own algorithm. The younger code monkeys need some milk. Write an algorithm that will instruct them exactly how to get milk from the fridge and pour it into a cup.

To test your algorithm, try reading it to a friend. Your friend must do exactly as the algorithm says, and nothing else. Did the algorithm work as it should? Could you make any changes to improve it?

Turn this page upside down to read the example algorithm below.

Example

Put an empty cup on a flat surface
Walk to the fridge
Open the fridge door
Take the container of milk out
Close the fridge door
Open the container of milk
Pour the milk into cup
Stop pouring when the cup is full
Close the container of milk
Open the fridge door
Put the container of milk back inside
Close the fridge door

Glossary

animated Made from many images put together in a way that looks like they are moving, such as in a cartoon

binary A simple language made up of ones and zeroes

complicated Made of many different parts and difficult to understand

compress In computers, to make a file smaller so it takes up less space and is easier to send from one computer to another

device A machine or piece of equipment designed to complete a certain task

GPS Short for Global Positioning System, a system that uses satellites to find exact locations

instructions A set of steps to be followed in order to complete a task

programs Collections of algorithms that tell computers to do certain tasks

Python A type of programming language that is made to be easy to use

satellites Human-made devices that travel around planets or other space objects to take photographs and collect information

Scratch A type of programming language made up of blocks. Scratch is a good programming language for beginner coders.

search results In computers, the websites that come up after typing words into a search engine, such as Google

statements Sentences that say facts or claim something is true

symbols Objects or pictures that represent, or stand for, something else

translate To change from one language into another

Index